by Maria Alaina

Table of Contents

Wholes 2
Halves.................................... 6
Thirds 10
Fourths 12
Parts of a Whole.................... 14

Consultant:
Adria F. Klein, Ph.D.
California State University, San Bernardino

capstone
classroom
Heinemann Raintree • Red Brick Learning
division of Capstone

Wholes

You can eat an orange.
You can eat the whole thing.

You can eat parts of an orange.
When you eat the whole orange,
you eat all the parts.

You can eat a whole apple.
This is a whole apple.

You can eat parts of an apple.
When you eat a whole apple,
you eat all the parts.

Halves

You can drink a whole glass of milk.

You can also drink half of the glass.
Half of this glass is full
and half is empty.

You can eat a whole sandwich.

This sandwich is cut in half.
The two parts are the same size.
Each part is one half.
Two halves make a whole.

Thirds

This is a whole pizza.
Three friends want
to share the pizza.

They can cut the pizza
into three equal parts.
Each part is called one third.
Three thirds make a whole.

Fourths

These four friends want
to share one apple.

They can cut the apple
into four equal parts.
Each part is called one fourth.
Four fourths make a whole.

Parts of a Whole

This pear has been cut
into two equal parts.
It is cut into halves.

This sandwich has been
cut into three equal parts.
It is cut into thirds.

You can share easily when
you split things into many parts.
Halves, thirds and fourths
are all parts of a whole.